WHISKERS AND BEAR

TRUE STORIES FROM AN ANIMAL SANCTUARY

GIACOMO GIAMMATTEO

INFERNO PUBLISHING COMPANY

Inferno Publishing Company

Houston, TX

For more information about this book visit my website.

Edition ISBNs

Trade Paperback 978-1-940313-38-2

E-book 978-1-940313-37-5

Cover design by Natasha Brown

Book design by Giacomo Giammatteo

This edition was prepared by Giacomo Giammatteo gg@giacomog.com

ISBN: 978-1-940313-37-5

❅ Created with Vellum

AUTHOR'S NOTE

If you've already bought the book, you can skip this note. If you're reading this at a bookstore or an online preview, continue on.

My wife and I have had an animal sanctuary for twenty-four years. We have always paid for everything ourselves, not because we're saints, but because we love the animals so much.

Two years ago; however, I had two heart attacks and two strokes, which forced me onto disability. Those unfortunate events have left us short on funds to keep things going. We still have about forty plus animals to take care of, and it would greatly help if we got a few additional funds. We're hoping this book will be the start, as all proceeds will go to the sanctuary.

I'm not begging for money; I'm asking if you'll buy the book. So how about it? Skip the cup of coffee today and buy the book. If you don't like books about animals, give it to someone who does.

INTRODUCTION

Most animals have a shorter life span than we do. It's sad, but true. That means any stories told will have happy *and* sad parts. Over the course of twenty-four years and dozens of animals, we've seen our share of sadness, but the overwhelming feelings and experiences have been ones of joy.

We not only love these animals; they make us laugh and bring us joy. This story is but one of many. It's about two dogs who lived their life to the fullest.

It's difficult to write about the beginning of something when you don't know the end. Or at least don't know *all* of the end. For now let's just say this is a story about two of the most magnificent dogs I've ever known.

Magnificent and crazy. Both of them.

ANOTHER GRAVE

I climbed up onto the tractor, a Kubota with a six-foot bucket on the front. It was a powerful machine, and we'd put it through the hoops more than a few times. What I mean is that my wife Mikki and I had dug a lot of graves.

I tied an old cloth diaper around my forehead and draped the end of it over the top of my bald head. There wasn't much better than a cotton cloth for keeping sweat out of your eyes, or the sun from burning your head. I turned the key and revved the engine. After letting it idle a moment, I lifted the bucket and drove toward the south side of the property where Mikki was waiting for me. She'd already gotten a few blankets and a clean sheet. For this one, she'd brought a pillow, too.

I reached up and wiped my eyes. I was getting damn tired of burying things.

An old white pickup crept down the gravel driveway, coming to a stop near the fence.

A neighbor leaned out and hollered. "What's goin' on?"

I wished he'd have kept going.

"Nothin'," I said, but not loud enough for him to hear.

The door opened, and he stepped out and walked over to the fence, using his right hand to shield his eyes from the sun as he peered over the top rail.

"What are you doin'?"

I could see there was no getting away from it. I muttered my answer a few times so my voice wouldn't crack when I yelled.

"Diggin' a grave," I hollered back.

"A grave? Which one died?"

Which one? That's what it had come to for most of the neighbors, and relatives, and friends. *Which one died?* As if it didn't matter. As if having forty-five animals made it easier to deal with when one of them died.

He came in through the side gate and headed in my direction. He walked slowly, which gave me time to compose myself. It's never easy to bury a friend, but this one…this one was special.

Mikki walked over to me. "He's just trying to help."

I nodded, acknowledging her wisdom, but all the while I thought…*I don't need his help.* Despite thinking that, the fact of the matter was I could probably use the help.

It hadn't rained in weeks, and the damn Texas ground was as hard as concrete. Even if the tractor *did* cut through, it could only go so deep; we'd have hand work to do at the bottom.

Our neighbor was about twenty feet away. He took off his hat and swiped at his forehead. It was a scorcher today and had been for a month or so.

"Who was it?" he asked.

I couldn't say, but I managed to gesture toward Mikki. She lifted the corner of the blanket so he could see.

"Oh shit!" he said. "I'm sorry."

"Thanks," I said.

He unbuttoned his shirt and grabbed a shovel that I had leaning against a small oak tree. "Might as well get this done."

I nodded again. He was right, of course, but I was in no hurry to put another friend in the ground. I cranked the engine up a little higher, shoved the tractor into low gear, and positioned the bucket for the first scoop of dirt. The bucket hit the ground with a metallic thud. It didn't do much more than break the surface.

"Whew!" the neighbor said. "Going to be a long day."

"That's for sure."

"How long have they been with you?" he asked.

They. I thought about what he said. I would have laughed if not for the circumstances. Everyone referred to the two of them as one. They or them. Bear and Whiskers. Whiskers and Bear. It was a cold day in July if anyone mentioned one without the other.

I handed him my bottle of water; he looked thirsty.

"They've been with us a long time. A *damn* long time."

A CHANCE MEETING

Twelve years earlier

ikki and I had been looking for property for a long time. We needed at least ten acres—or so we thought —but we also needed it to be close to a good hospital. We had been looking for months, to no avail. Everything we found was either too far away or too much money. One Saturday, after checking three different properties, we thought we found one that could work. The house came with eighteen acres and was the right price. It was almost thirty miles outside the city, but we were tired of searching, and this was the closest we'd seen to what we wanted. We told the owner we'd get back to him on Monday.

On the way home, we stopped to get a bottle of water. While I was at the corner store, I picked up one of the local Realtor magazines.

Mikki looked at the magazine as I got in the car. "We've already got a place."

"No harm in looking," I said. "We haven't made an offer yet."

While I drove, she leafed through the magazine. "Here's one that doesn't look bad—two miles from The Woodlands. Seven acres with a half-acre pond."

"Seven acres isn't enough."

"It's got a pond," she said. "The other place didn't. And this one's only two miles from The Woodlands."

View of house across the pond

I turned right at the next light so I could make a U-turn. "I guess we can take a look."

I called the Realtor and got directions, then we navigated our way through an old subdivision and down a dead-end street to a locked gate with a big Private Property sign. A pick-up was coming up the gravel driveway as we arrived. The driver stopped and rolled down his window.

"I'm here to look at the property," I said. "Are you the owner?"

He turned around and invited us to follow him. We weren't halfway down the drive before Mikki said, "Oh, my God, will you look at that?"

I looked and saw two Australian shepherds running alongside his truck. "Damn, they look just like ours."

Mikki was convinced it was a sign. After looking at the place, I couldn't argue. The house was a cute little ranch with a full-length front porch that overlooked the pond. Behind the house was a forty-by-sixty foot barn. It was perfect. Best of all, the property was surrounded by woods.

"This looks like more than seven acres," I said.

"It is," he said, "but I'm only selling part of the ground on the north side of the road. There are four more acres on the north side, and another six on the south side. And one hundred acres of woods sits beyond that."

"If you sell the four acres on the north side at a reasonable price, we'll take it," I said.

He and I haggled a little over price, but within a half an hour we negotiated a deal. One month later we made settlement.

Mikki and I decided to build an addition on the back of the house before we moved in. We also wanted to build a separate office for me with an apartment above it for our son.

While construction was going on, Mikki and I went to the property every night and weekend to build fences. It was during those first few months that we met Bear.

We were driving down the street toward our property when a large black dog bolted from the woods on the side of the road. I slammed on the brakes, afraid I would hit him. But he wasn't interested in being hit; he only wanted to bite our tires. And then he lunged at my window. I quickly pulled my arm in. He scared the hell out of me.

We continued on, the dog chasing us halfway down our driveway. "Hell of a way to be greeted," I said.

"Somebody needs to keep that dog locked up," Mikki said.

The next two trips to the property were a repeat, with the dog coming at us from one side of the street or another. I tried to get out one time, but he would have nothing of it, so the next night I brought treats—a few steak bones and a couple of dog biscuits.

The dog lunged again, but this time I stopped the van and tossed a small biscuit into the street. He gobbled it up. Then I reached out with the steak bone. To my surprise he took it from me gently. We continued to meet the dog every night that week. After about three days, the attacks stopped. He just sat on the side of the street waiting for us. This time Mikki got out and fed him by hand.

One of the neighbors came out and said, "I see you've met Bear."

We laughed. "Is he yours?" I asked.

She shook her head. "Bear doesn't belong to anybody. Two people have tried keeping him, and he won't have it. He lives in the woods. And in case you haven't noticed, he *owns* this street."

We soon discovered that truer words had never been spoken. Bear *did* own that street, and he let everyone know it—cars, vans, trucks (especially delivery trucks), but most of all—other dogs. If another male dog ventured down that street he was in for trouble. Bear once fought a Great Dane, a huge old boy who weighed about 170 pounds. The dog put Bear down three times in less than a few minutes, but Bear got up fighting each time. We were finally able to pull him away from the fight, but he didn't like it one bit.

We continued paying Bear's biscuit toll, as we liked to call it, every time we came down his street, and he became friendlier each week. By the time we finished construction in August, Bear had a new friend, a female dog who had wandered up and claimed the house at the end of the street.

The homeowners came out one day and found her lying on their porch. Two months later she was still there. They named her Princess. It didn't take long to figure out that she was anything but a princess.

Together, she and Bear made a formidable pair. They double-teamed all vehicles, strangers, and—God forbid—other dogs who ventured onto Springwood Drive, at least below the 700 block. And we soon discovered their claim to territory didn't stop at the street.

We moved into our house on a Friday, bringing our three dogs with us. Bear and Princess weren't in the street when we passed, but less than an hour later Bear marched down our driveway, Princess a few feet behind him. I had no doubt as to why he was there. He wanted to introduce himself to our dogs and let them know the rules.

I said he marched down the drive, but marched wasn't the right word. *Strutted* was more like it. Bear had a strut that let everyone know who he was and what rank he held in the neighborhood. At first I thought he was trying to let other dogs know he was boss. Now I know better; it was to let people know he was boss; the dogs already knew.

Our driveway is about 1,000 feet from the gate to our house. We didn't see Bear coming, but our dogs did. Flash, Kassie, and Kelly sniffed the air, then raced up the gravel drive. I knew there was going to be trouble.

By the time Mikki and I reached them, a fierce fight was in full swing —Bear and Princess against our three dogs.

We had broken up dog fights before. It's never an easy thing, but it's doable if you've got two people and two dogs; five dogs takes it to another level.

Not knowing what else to do, I hollered a few times. I waved my hands in the air and hollered more. And suddenly Bear stopped. Princess followed his lead. Our dogs continued their aggressive behavior, but once Bear stopped, we grabbed ours and yanked them back. Bear sat on the drive, five feet away, as if nothing had happened. It was just another day in paradise.

"Did you see that?" I said to Mikki.

"I didn't think we'd stop it," she said.

"We didn't stop it," I said. "Bear did."

Bear came to visit every day after that, and Princess was usually with him. Our dogs still hated Bear, but we had the fence up on the back section now, and our dogs stayed behind it. Bear didn't seem to care one bit about them. He didn't try to fight with them; he didn't growl at them. He didn't even seem to dislike them. He just came to visit us.

Bear wasn't much of a conversationalist. He'd meander up, greet us, and then find a shady place to lie down. On hot days he'd often take a swim in the pond and then come back for the shade, still dripping wet. Most dogs shake off after a swim. Not Bear.

He'd stay for an hour or so. Mikki and I talked to him while we worked, for no reason other than it gave us something to do. When he was ready, he'd get up and head out, but never in a hurry. One day, though, I noticed his head perk up, and he sniffed the air. Then he bolted toward the driveway.

I put the tools down and took off after him. "Something's up," I said to Mikki.

ARCH ENEMY

*S*omething was a large black dog who appeared to be at least part Great Dane. He towered over Bear and looked to weigh about one hundred seventy pounds. To put it in perspective, Bear weighed fifty-five pounds.

It was better for all involved that this dog was still immature. He was big, but he hadn't developed his fighting skills, or the heart for it, I suspect. After a short tussle, the dog submitted.

I learned something else about Bear that day—he was never out to hurt another dog, just to show it who was boss and get it to acknowledge him. After that he was fine. He and "Duke," as we came to call him, became friends, but the friendship only lasted a few days—until we found Duke's owners.

Two weeks later, Duke showed up again, and once again Bear forced him to submit. The third time Duke came around, the fight lasted much longer. Duke was gaining experience and confidence. This fight drew blood, and most of it belonged to Duke. We took him to the vet and got him fixed up, then drove him back to his owner. "Please try to contain him," I said, and left it at that.

We had finished building the fence in the back of the property and started enclosing the seven acres in the front. It was an arduous task, and Mikki and I were getting tired of it, though it was rewarding to see each section finished.

The other rewarding thing was that every day Bear and Princess came for their visit. It was during these strolls to our property that I noticed Bear stop and sniff certain sections of the driveway and then take off into the woods. One day, shortly after he took off, we discovered why he did it.

I was working the fence about midway down the property when I heard Mikki holler, "Watch out!"

I looked up. A coyote was racing toward me, but right on his heels was Bear. Coyotes are fast, but Bear kept stride. He chased that thing across the property—about 500 feet—and into the woods on the other side. Twenty minutes later, Bear came back, went to the pond and took a drink, then went for a swim. After his swim, he strutted over and plopped down in a sunny spot on the grass about twenty feet from me. There was a look on his face, a smile that said it all. He'd done his job.

Bear in grass

It was that day I realized our property had fallen under Bear's protection. And so had we. Our dogs didn't matter to him; they could do what they wanted. But we were being protected by Bear. I laughed about it, but inside I got a nice, warm feeling. Oddly enough, I felt a little safer.

ANOTHER FRIEND

*S*ometime in October I heard Kassie and Flash going nuts, barking like crazy. They did it every time Bear came down the drive, but today it sounded more fierce than normal. I walked down the drive to meet Bear and Princess. They had another dog with them, a little puppy with coarse, tangled fur. She was adorable. Bear greeted us, then stepped back and let us make a fuss over the pup. After that, he turned around and left, as if he'd only come to introduce us to his new friend.

We found out later that the new pup's name was Whiskers, and she belonged to a family that had just moved into the neighborhood. Before long the family realized that while they may have *technically* owned Whiskers, she really belonged to Bear. Within weeks she stopped going inside their house, choosing to spend her nights roaming the woods with Bear and Princess.

Even that routine wasn't to last. When Whiskers was about six months old, Princess stopped going out at night. A few months after that she delivered eight adorable puppies. Mikki and I swore we wouldn't take any of them, but when six weeks passed and there was still one left, we gave in. It was a male, and we named him Slick, but his story is for another book. Here's a picture of me on the front porch, holding Slick. It's the picture I use for my author bio.

Giacomo and Slick

Princess became domesticated after that. She still hung out with Bear on the porch or in the front yard, and she played with Whiskers in the front yard, but she didn't go out to roam the woods anymore; she left that up to Whiskers. Bear wasn't about to go out with a novice, though, and he took it upon himself to train the pup.

Those two were as much wild dogs as anything I've seen. No one ever fed them. They caught their own food—rabbits, voles, mice, and lots of squirrels. And in late August we learned they had expanded their diet.

ANIMAL CONTROL

*M*y daughter-in-law, Missy, was at my house helping me move a treadmill into the garage. Bear and Whiskers were with us when a truck pulled up.

"It's Animal Control," Missy said.

We walked out to the driveway to see what they wanted. The driver rolled down his window and said, "We're lookin' for two dogs that got a few chickens at a place about half a mile from here. According to the owner, one of them is black and the other's a mangy-looking brown cur."

When I didn't say anything, he said, "You seen 'em?"

From the corner of my eye, I noticed Missy scooting the dogs inside the garage, trying her best to hide them. I moved to block the guy's view. "You say they stole his chickens?"

"Right out of his back yard. And he said it wasn't the first time."

These weren't my dogs, and I had no business poking my nose into the situation, but I had a lot of respect for Bear and Whiskers, and it would be a cold day in hell before I surrendered them to Animal

Control. "We'll keep an eye out for them," I said, and tapped on the side of his truck, as if to send him off.

He handed me a business card and drove off. I went to the garage where Missy had secured the dogs. "This is bad," I said. "Somebody's going to shoot those damn dogs."

I told the people who owned Princess, which was where Bear still rested during the day, and I mentioned it to the family who owned Whiskers, but neither family seemed to care; in fact, the ones who owned Princess said they were about ready to call the SPCA (Society for the Prevention of Cruelty to Animals) on Bear. He had bitten one of their guests and yet another UPS driver.

I got a little riled. "Don't *dare* call the SPCA. If you don't want him, I'll take him."

I didn't know what I was going to do with Bear, especially considering my dogs hated him so much, but I'd figure something out. There was no way that dog was going to the SPCA.

Bear must have been listening and wanted out of there, because the very next week he bit a FedEx driver.

We got the call around dinner time. "He's done it again," the neighbor said. "We can't keep him any longer."

I picked Bear up, though that was no more than opening the side door of the van and letting him jump in. My son and his family had recently moved into the neighborhood, so I asked them if they'd be Bear's caretakers, because that's about all it amounted to.

And so Bear went to live with Jimmy, but Whiskers stayed under her bridge on the street—the troll of the neighborhood. To Bear and Whiskers, this was nothing but a small glitch; it never interrupted their nightly routine.

I've come to believe that dogs know a lot more about us than we do them. Evidence of this showed again when Bear exhibited a knowledge of his transition. Somehow he knew that he was no longer welcome at Princess' house.

He would visit her when he stopped to pick up Whiskers on his nightly run, but he didn't try sleeping at Princess' house anymore, and he never bit anyone at their house again, not even the dreaded delivery men. He was content to live in Jimmy's house about four hours a day, something he had never done before. The only difference was, once it got dark he demanded to be let out, and he wouldn't come back until morning.

Bear immediately adopted Jimmy's family, and Jimmy's kids fell under his protection. Bear wouldn't let anyone go near them. He was a near-perfect dog while he was in the house—if you ignored his sleeping habits—but he acted as if he owned the house (like he did everything). He slept where he wanted and did what he wanted. Below is a picture of one of Bear's typical sleeping places.

Bear on Hockey Table

It's not like this picture was an unusual catch with the camera. Bear could be found resting or sleeping almost anywhere, including coffee tables, end tables, and even kitchen tables. He had his own rules. Below are a few of the pictures.

Bear on coffee table

Bear on end table

Bear on sofa table

The crazy thing was this—he would let me and my wife or my son and his wife do *some* things with him, but he would let my grandkids do almost anything, including dress him up with bunny ears, put glasses on him, antlers, hats, and more. And he acted as if he were having a good time while doing it.

KING OF THE STREETS

*A*bout two weeks after Bear moved in with Jimmy, the people who owned Whiskers packed up and moved, abandoning her to the streets. They didn't even say goodbye to her. The good thing was that Whiskers didn't seem to mind.

It soon became obvious to the neighborhood that Whiskers was Bear's new sidekick. She belonged with Bear, and everyone knew it. She hung out with him all the time, and now that her owners were gone, it was official.

Whiskers and Bear became inseparable after that. But for all that Bear dominated every animal and person in the neighborhood, he had a more difficult time with Whiskers. Maybe he'd trained her too well, or maybe she just didn't give a darn. Either way, Whiskers had a mind of her own.

Six months passed, and the Bear-and-Whiskers team showed no signs of letting up. They were a perfect team for hunting and fighting. Whiskers was the decoy when it came to fighting, drawing attention by using feints and charges. Bear did the heavy work, taking on all comers, no matter their size or species. They fought invading dogs from other parts of the neighborhood, they fought coyotes, and they even fended off wild boars, though they were smart enough to stop short of physically engaging them.

Hunting was a different strategy and depended on the quarry. For squirrels—once they treed one in a spot with no escape options—Whiskers sat and waited. Bear would make a big fuss and then leave, as if he had given up. Meanwhile, Whiskers lay camouflaged in the tall grass or in a bed of leaves.

I often watched, but even when I knew where Whiskers was, it was difficult to see her. She'd sit as still as a rock for hours on end and wait for the squirrels to come down. She got about half the ones they treed. Once she caught them, Bear would take them from her, and then they'd share the meal, but he got his portion first.

For other prey, like rabbits, Whiskers did the chasing while Bear positioned himself for an ambush. It was amazing to watch them work and to witness the communication between these animals. They had a plan. They *talked* to each other. They knew what was expected of each other. And they delivered.

COMMUNICATION

I once saw Whiskers run out of the woods to where Bear stood, by the drive. She let out barks and other sounds I didn't recognize, and then she went back into the woods the way she had come out. Bear took off down the drive at a ninety-degree angle from her. He ran about fifty yards, then stopped and waited in a small grove of trees. Within half a minute, a small deer ran out, Whiskers hot on its tail. Bear bolted to cut it off. Between the two of them, they almost got it. The deer broke free and leaped the fence where I was working. Bear jumped and gave chase but the deer had too big a lead. It was safe.

I had to replay the scene in my head a few times before I told anyone. I almost couldn't believe what I'd seen. This was hunting at a sophisticated level.

I'm going to make a side note here and talk about animal communication. We humans tend to look at animals only from our own perspective. In order to measure intelligence, the animals must meet *our* standards, must perform to the tests *we* develop. But suppose we look at animals on a different level, from their perspective. Suppose we examine them based on what they do and how they perform.

The bottom line is we see animals as *they* want us to see them. If we have one or two dogs or cats, they react and perform as we want them to. But when you examine a pack with a chance to roam and be themselves, you have a different scenario.

We have fifteen dogs, enough to qualify as a good-sized pack. And these dogs *are* a pack. They *pretend* to listen to us, *pretend* to need us, but they operate on their own. I have no illusions about who's boss—Bear rules the roost. He rules it even when he's not here. But if Bear rules, Whiskers is his right-hand man, or in this case, his "right-hand woman." She tells the other dogs what to do when Bear is away. It's amazing to watch.

Another time I witnessed this communication, Whiskers was inside the fence on the seven-acre part of our property. I heard her unique bark/howl from my office and went out to see what was going on. Bear was just coming down the driveway. Whiskers ran to meet him at the fence. She "talked" to him for about ten seconds, and then he jumped the fence and ran toward the back of the property. Intrigued, I walked outside and continued to watch.

Whiskers then ran to the back of the house and returned with two other dogs—Biscotti and Mollie. She led them to a small wooded area. Within seconds, they scared up three rabbits and herded them toward Bear.

I hated to see it happen, but Bear caught one of the rabbits. And when the other two turned around, Biscotti caught another. The third one got away. The amazing part of all this was the coordination and communication that took place between the dogs. I have no explanation for it.

PAYING TRIBUTE

*L*ate in the summer, a young lady bought the house next to my son's. It had been built about six months before, and she was its first owner. Two or three weeks after moving in, she came to my son's house one evening.

"You're the one who owns the black dog, right?"

I'm sure he had the same *uh-oh* type thoughts I've had a few times. "Bear lives here," he said. "Why? Did he cause trouble?"

"Not really trouble," she said, and then paused before saying "I don't know how to ask this, but do you feed him?"

"Of course we feed him," Jimmy said, trying to keep his indignant expression in check. "Why?"

"Do you know that he comes by my house every night?"

"What do you mean?"

The neighbor shifted her weight. "Every night around eleven, he comes knocking on my door." She laughed, then said, "I mean actually knocking. The first time I heard it, I thought it was a person; now I recognize his special scratch."

Jimmy looked puzzled, I'm sure. "What does he do?"

"The first time he just stood there. I was eating a hot dog at the time, so I broke off a piece and gave it to him. He's been back every night since." She laughed again. "Now when I answer the door, he just walks in as if he's been invited. Sometimes he goes right to the kitchen and sits in front of the refrigerator, and sometimes he goes to the family room and gets on the couch, like I invited him to watch TV."

"God, I'm sorry," Jimmy said. "I'll—"

She waved her hand. "No, don't worry. I'm not complaining. It's actually kind of nice. One night I had one of the pilots staying with me, and Bear growled and scared the hell out of the guy. It kind of made me feel safe."

"Are you sure?" Jimmy said.

"Positive. I just wanted to make sure he wasn't starving, although he didn't look like it. Now he comes by and gets his treat, stays a few minutes and then leaves. It's really cute."

In talking to other neighbors, we discovered Bear had a routine. It was like a paper route, only it was a food route. There were six houses he went to every night, from 8:00 p.m. to almost midnight. One house gave him cookies, another biscuits, and a third table scraps. At the other *stops* he got whatever was available.

No matter what, the neighbors all said Bear's routine was the same—at each house he'd walk up, scratch the front door, and then either get his treat there or go inside for a visit. When he was done, he'd leave. If it hadn't been for the flight attendant, we might never have known."

BUT THAT WASN'T ALL BEAR DID

*W*e soon found out more about our friend Bear. Not long after this incident, a lady pulled up our driveway in a big SUV.

She rolled down the window and shot me a penetrating glare. "Do you own a dog named Bear?"

Uh -oh. I walked toward her car and said, "I guess as much as anyone can own Bear—yeah, I do."

She got out of her car, walked to the back and popped the rear hatch. Inside was a box of puppies—nine of them. The question must have been written all over my face, because she didn't hesitate to provide an answer.

"A few months back Bear jumped our six-foot fence, fought off our pure-bred male boxer, and mated with our pure-bred boxer bitch." She then shot me another wicked look. "Pick one," she said. "It's the least you can do considering what that dog did."

She had a little attitude in her voice, although I can't say I blamed her. Her heart had been set on breeding boxers, not delinquent mutts.

She set the box on the ground and the puppies crawled out. Most of them cuddled with each other. A few hid under the car. But one of them—a brown one that looked all boxer—walked up to me. She didn't shy away when I reached for her; instead she gave me a tiny snarl.

Some of the other puppies looked similar to Bear—black with a few white markings—and a few of them were males, but the bold female was the one I chose. She didn't look a bit like Bear, but she *was* Bear, right down to the confident strut.

Mollie

"I'll take *her*," I said.

Now all I had to do was find a way to tell Mikki we had increased our dog count by one. I held the dog in my arms and inhaled her puppy breath as I walked to the house. About an hour later we gave her a name—Mollie.

Now she was official. Once an animal got a name, they had a home.

A BAD SIGN

$\mathcal{A}$bout six months later, on my way to Home Depot, I stopped on the street to say hi to Whiskers. She was under her bridge, as usual, but I couldn't get her to come out so I went on my way. Later that night she was still in the same place. Bear was sitting by her side.

I got out of the van and went to her. She still didn't come out. When I reached in to pet her, she pulled back. I knew something was wrong. I called my wife to come help. When Mikki arrived, we pulled Whiskers out from under the bridge and onto a blanket. She couldn't move her back legs. We loaded her into the van and took her to a vet. The vet suspected a disc problem and didn't offer much encouragement.

"What are we going to do?" Mikki asked. "We can't afford to pay for this."

It didn't take us long to decide. One look into Whiskers' eyes, and we knew we had to help. I asked the vet if he'd work with us on payments. Whiskers stayed for almost three weeks, but she didn't make much progress. The vet recommended we put her down. There was no way we were doing that, so we asked for anti-inflammatory pills, and we took her home.

We couldn't take her to our house because our dogs still hated Whiskers and Bear, so my son Jimmy agreed to take her in. He only had one older dog, who we presumed would be no trouble, and of course, Bear.

For three weeks, Jimmy and his wife, Missy, nursed Whiskers. He carried her out every day to go to the bathroom, and he "walked" her by holding her back legs up using a cloth strap, which allowed her to use her front legs. Still, there seemed to be no progress.

The following Monday, he heard a slight whimpering noise at the door. When he looked over, he saw Whiskers, standing on her own and wanting to go out. He opened the door and she walked out, into the street, and all the way down to our house, about a quarter mile away.

Jimmy called to tell me that Whiskers was headed toward my house. I watched the drive until I saw her. She stopped every forty or fifty feet to rest, and then she continued at a slow pace in our direction. When she got close to our back door, she scratched out a small spot in the dirt near the porch and lay down.

Whiskers sleeping in dirt

I tried getting her to come inside but she wanted nothing to do with it. Our dogs were going crazy, barking and scratching at the fence. Whiskers paid them no mind.

Mikki came outside to see what the commotion was about. She looked at Whiskers and said, "What's she doing here?"

"I guess she came for a visit," I said.

Later that night Whiskers dug up a bush from Mikki's garden, one that had been close to the fountain, and dragged it away using her teeth. After that she worked until she got the hole just right then lay down to sleep. "I guess she's here to stay," Mikki said.

ESCAPE

It took Whiskers another week to get back to normal. By that time we had almost seven acres fenced in, and our dogs seemed to have gotten used to her, so we put Whiskers in with them.

She didn't like it one bit.

I watched her from my office window. That first week she spent the better part of each day looking for a way out. She walked the fence line, dug in numerous spots, and tried climbing the fence, but she couldn't escape. I thought we were good—until Friday.

I heard the UPS driver beeping his horn, and when I looked, I saw Whiskers in the middle of our driveway, blocking his truck. I got her and put her back inside the fence.

For the next two days I tried to discover how she'd gotten out, but I couldn't. She'd start out the day inside the fence, then all of a sudden she'd just be gone. I walked the fence looking for holes and found nothing.

On the third day I figured it out. She was about halfway up the driveway, inside the fence. She looked around, as if checking to see if I was watching, and then she walked into a drainage creek and disappeared. I followed and found her escape route—a small drainage pipe that led from the ditch, under the driveway, and into a ditch on the other side. The pipe was barely bigger than Whiskers, and she had to crawl about sixty feet to get to the other side, but to her that didn't matter. All that mattered was getting out.

We sealed the hole, but within two days she was out again. This time we found a fallen tree she was using to get over the fence. We thought we had won the battle after we cleared the tree, but she soon started an all-out excavation campaign.

She dug a hole so deep, that she tunneled under the hog panel and crawled out in a feat that a contortionist would have been proud of. After witnessing that bit of defiance and determination, we decided to let her run free.

Whiskers in hole by fence

To our surprise, she stayed at our place, and occasionally even came inside to sleep, but she refused to be fenced in. The new arrangement worked perfectly. Every night, long after dark, Bear trotted down the driveway to "pick her up" for his date, and the two went hunting.

HUNTERS AND HUNTED

I believe Bear and Whiskers hunted just about everything. I can't swear to it, but it seemed as if they had no fear, and sometimes, I think they lacked common sense. They fought anything that came near our property, and they hunted anything they could eat. They seldom ate at home, choosing instead to live off the woods they loved roaming in so much.

Squirrels, rabbits, guinea hens, chickens, voles...Bear and Whiskers didn't seem to be selective. We often found Whiskers or Bear munching on a bone or piece of meat that we had no idea where it came from or what it was, as was the case with the picture below (or the next page).

Whiskers with rib cage

I don't know if Bear and Whiskers ever got a deer, but several times we found remains that looked as if they did. One time Bear came back with a baby wild pig. I never told my wife about that; she'd have been mortified, as our animal sanctuary had started with the rescue of pigs.

But it was a testament to his bravery (or stupidity), his prowess, and the teamwork of Bear and Whiskers that they managed to get a pig. I can attest to the adage about nothing being more dangerous than a mother pig guarding her young. They are formidable and fierce.

One night, though, Bear tackled something he shouldn't have.

We were under a severe weather watch, with heavy thunderstorms and the threat of tornadoes. I lured Whiskers inside and kept her there, despite her protests to go out. My son didn't fare as well with Bear. He gave the signs of needing to go out and do his business but as soon as Jimmy opened the door, Bear bolted.

Twice that night Jimmy called, asking if I had seen Bear. I hadn't, and it was far too nasty to go into the woods looking for him. The next morning, my daughter-in-law went to the door to get Bear. The porch was covered in blood.

Blood was all over the concrete porch and the sidewalk. Panicked, she followed the trail of it to the driveway, where she found a pool of blood next to the car. Bear was underneath. He limped out, but it didn't look good. He was bleeding from his back, his side, his stomach, shoulders and neck, and his face was mangled. Most of the blood was coming from his eye, though, which was a mass of blood.

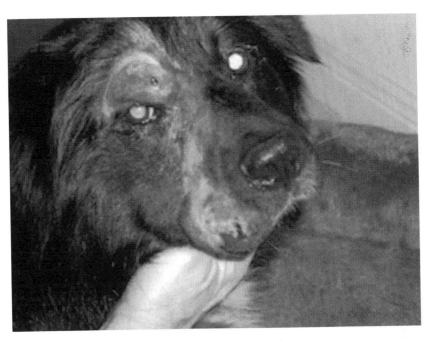

Bear after a run-in with raccoons

Jimmy and Missy rushed him to the vet's office and after a few hours he was stitched up, sewed up, medicated, and sent home. The vet said he felt as if it was at least one, if not several, raccoons, and that Bear was lucky to still have his eye. Bear wasn't "out of the woods" so to speak, but the vet thought he would make it.

I would like to think that Bear learned his lesson that night—that there are some things on dark stormy nights you don't attack—but I don't think he did. What he may have learned, though, is that he shouldn't attack anything without his sidekick.

GENERAL WHISKERS

*P*ound for pound, Bear was one of the toughest dogs I've ever seen. Despite that, before he paired up with Whiskers, he often came home bloodied from battles with God knows what. The run-in with the raccoon was the first time he returned with even a scratch since he'd been going out with Whiskers, but she hadn't been with him that night.

I got to thinking about that afterward, and I realized that although Bear was a courageous, relentless fighter and a skilled tactician, Whiskers was the strategist and general. I came to this conclusion after mentally replaying many of their incidents.

- When they got the guinea hens, it was Whiskers who circled behind them and drove them toward Bear.
- When they almost got the deer, it was Whiskers who drove it from the woods.
- Whiskers who flushed the rabbits from their haven in the thicket.
- Whiskers who lay in wait for the squirrels.

- And it was Whiskers who always flanked whatever they were fighting, or acted as the decoy so Bear had a clear approach.

This was an enlightening moment for me. I realized Whiskers truly did communicate with the other animals, and that these animals *talked* to each other. It's a shame I didn't take the time to record the different scenarios, or at least to record the various sounds they made with their barking, howling, yipping and such, because I'm convinced each one had a special meaning.

I noticed the way the dogs reacted differently to each sound. Sometimes they perked their ears up but did nothing. At other times they went on alert but still not much reaction. And there were some barks that sent the dogs into a frenzy, where they began barking and demanding to go out, all racing to the door at once. On a few occasions, I went outside to try and find out what was going on. On at least two of the "high-alert" incidents, there was real danger. Once, two coyotes were in the side yard, and another time there was a wild boar. It taught *me* to listen to Whiskers as well.

If you're at all interested in animal communication, look up John J. Pratt. He's done some phenomenal work recording the language and communication of prairie dogs, and it is nothing short of amazing. It's worth it to take some time and check out his site. http://www.john-pratt.com. You'll never think about animals the same way again.

THE WILD-BOAR INCIDENT

*W*hiskers had been with us for several years, and in all that time, she'd never left the property to roam the neighborhood. She ventured into the woods or into the wooded section of the property on the other side of us, but she never returned to her old stomping grounds. And after dark, unless she was hunting with Bear, she stayed north of our gravel drive. It was as if an invisible line marked the territory.

One night in late November, Whiskers rushed toward the house, barking her crazy hoot-owl bark. It must have been a high-alert call because the dogs inside went nuts. I let the other dogs out but they were restricted to inside the fence. Whiskers raced to the driveway, stopping about halfway up where the trees were thickest.

I went out to see what was going on. Before I got to where she was, I knew something was up. Whiskers was in rare form and unusually agitated, but she wouldn't go into the woods. The other dogs, also extremely agitated, were inside the fence but doing their best to get out. As I drew close to Whiskers, I heard a ferocious grunting-growling sound, and recognized it instantly—a wild boar.

I grabbed hold of the scruff of Whiskers' neck, and gave her a light tug. "Come on, Whisk. You don't want any part of that," I said.

But she wouldn't listen. I tried a few more times, and then she took off running up the drive. Within minutes I got a call from my son.

"Whiskers just came banging on our door," he said. "And Bear was going crazy."

"Did you let Bear out?" I asked.

"I had to. He was flipping out."

Within a few seconds, Bear and Whiskers came racing toward me, and this time they didn't hold the line at the driveway. Bear rushed into the woods on one side, Whiskers on the other. I heard a few more grunts from the boar, and then it went off into the woods. Bear and Whiskers didn't follow.

I was thrilled that the boar left without a fight, as it might have done a lot of damage to the dogs, and I was impressed those dogs were brave enough, or crazy enough, to drive the boar away, but I was more impressed that Whiskers had the sense to go get Bear for help. That displayed an intelligence I didn't think dogs possessed, at least before we got Whiskers.

INTELLIGENCE

*A*s mentioned before, dogs are usually judged as being smart or not smart based on a set of standards designed by people. We're amazed by the dogs who follow orders, and obey commands. Look at any dog show or exhibition of dog intelligence and you'll see demonstrations where dogs follow human commands—sitting, lying down, prancing, dancing, speaking, counting—the list goes on forever.

Whiskers never did any of this. There wasn't a single command she would obey. She wouldn't sit, speak, heel, come, stay, or any such nonsense. In fact, Whiskers did nothing a normal dog would do. She didn't even follow pack behavior.

She wouldn't eat with the other dogs or sleep with them. She wouldn't sleep inside the house, and when she did come inside on occasion to eat, she wouldn't drink inside, opting to go outside and drink from the fountain. She also had an unnatural aversion to pictures. I swear! If ever I pulled out a camera, Whiskers was gone. It's why I have so few pictures of her.

But for all of her lack of intelligence she was brilliant. A story might give the best example.

One year, a housing developer bought the forty acres north of us and soon cleared the woods. It drove the wildlife into the only wooded areas left: the few acres we had and the roughly one hundred acres south of us. I'm sure that made for crowded conditions, as there were already plenty of coyotes and wild boars and pigs living there.

It wasn't long before Whiskers felt the pressure. Coyotes began encroaching farther onto our property, and they were getting bolder. We found more droppings on the driveway every week. The driveway seemed to be the line in the sand, the final battleground, and that's where Whiskers took her stand.

One night, after a bad storm which forced Whiskers inside, she woke me early by banging at the door and howling. I let her out, and she raced toward the waterfalls. When I looked closer, there was a coyote atop the waterfalls, just sitting. When it saw Whiskers coming, it got up and left, but she must have been appalled that it had the nerve to invade *her* property. I still don't know how she knew it was there. Below (probably next page) is a picture, though not a good one.

Every night, after patrolling with Bear, she would stake out a spot not far from our house, where there was a small patch of land between the barn and the driveway. No matter what, she refused to surrender that territory. One night I heard a horrible ruckus, and she came running to the door. I tried getting her inside, but she didn't want that; she was coming for help.

When I realized it, I let Biscotti outside and the two of them raced back toward the patch of ground. I followed. By the time I got there, the fight was in full swing—Whiskers and Biscotti were fighting off three coyotes. The coyotes fled quickly when they saw me.

After that, things got worse. Every night about 11:00 p.m., a train passed by about a half a mile from our house. When the whistle blew, it spurred the coyotes to howl. It was a beautiful awe-inspiring sound, but I worried because they had started moving in closer. We suspected we had lost one of our old dogs to them a few years back, and now we had a feral cat who had decided to make her home at our place. She gave birth to five kittens soon afterward.

I had no reason to worry though.

Upon hearing their howling and Whiskers' alerts, I went outside with a light. A few of those times the coyotes were ten or twenty yards on the other side of the drive, but Whiskers kept them at bay, never letting them come past the driveway. On the nights when Bear was with her, they wouldn't even come close. I think word about Bear may have spread through their pack.

BACK TO THE STORY

*T*o continue the story, my son and his family went on vacation, and Bear was confined to the house. We watched him while they were away. We let him out during the day but kept him inside at night. It was on one of these nights the coyotes decided to push the limits. They got past Whiskers and were close to the barn, where the cats were.

We managed to chase them away, but it's what happened afterward that surprised and impressed me. Whiskers climbed atop a flat-bed trailer we had parked in the side yard, and she stayed there, refusing to move. That night, Whiskers slept on the trailer by the barn instead of her usual spot. I'm convinced she was protecting the cats, even though she never seemed particularly fond of them.

Whiskers on trailer

The next morning she began a major excavation project, digging a huge hole next to the fence by the side of our house. At first we thought she wanted a new place to lie down, as this was her standard procedure, especially on hot days. She'd dig a foot or so into the ground and then go to sleep in the cool dirt. But we quickly realized this was no ordinary sleeping hole. This hole was far too deep.

By the end of the second day, the hole was more than two feet deep and almost three feet across. It was now below the fencing, and it was obvious she was trying to break through. My wife said, "She's trying to get inside the fence."

I looked at her and shook my head. "No way Whiskers wants inside the fence. She won't go there." Then it hit me. "She's trying to bust the other dogs out!" Sure enough. That's what it was.

There was no reason for Whiskers to dig that hole other than to break the other dogs out. She had proven time and again that she wanted no part of the fenced-in area and, in fact, wouldn't tolerate it, so why else would she be digging a hole? And why did she *sleep* in it? Was it to cover up her activities? The picture below shows the beginning of her digging.

Whiskers at beginning of dig

We tried putting her back inside the fence to test the theory, but she would have none of it. And after we covered up her hole, she dug another one. Finally, I had to foolproof the fence by placing metal rods in the ground so she couldn't dig.

I'm convinced that she wanted the other dogs to come out at night to help her guard the property. We tried bringing her in at nights so she'd be safe, but she wouldn't have any of that either. Whiskers had a mission, and nothing was going to deter her.

COPPERHEADS

*O*hen I said before that Whiskers wouldn't let anything on the property, even I didn't fully realize how far she was willing to take that commitment. She ferociously guarded our property against wild boars and coyotes. Stray dogs didn't dare attempt to enter the property, nor cats. Squirrels soon learned to keep their distance, and even the biggest of the blue herons would take flight when Whiskers gave chase.

But it wasn't until a scorching hot day in August that I discovered that the line she drew in the sand included snakes.

We had a lot of snakes on our property, including what seemed like more than our fair share of copperheads and a few water moccasins. When we spotted them, either my wife or I would catch them and relocate them to a spot deep in the woods, a few miles away. Whiskers, however, wasn't so accommodating.

Mikki was working in the garden when she noticed Whiskers coming up the sidewalk walking slower than usual, which meant very slow. "Hey, Whiskers," she said, but Whisk didn't even look at Mikki; instead, she went to the fountain and drank water and then walked toward the back door. It was then that Mikki noticed her face seemed swollen.

Mikki went to let her inside and, upon inspection, realized what had happened—Whiskers had been bitten. Mikki called me on the phone. "I think Whiskers was bitten by a snake," she said. "Her face is very swollen."

I came inside, and we tried tending to Whiskers, one of the most difficult things to do. She never let people mess with her. No combing. No shots. No medicine. Seldom did she tolerate being petted. She tried leaving the house several times but Mikki picked her up and laid her on a sofa we had in the living room. I grabbed a pillow to keep Whiskers' head raised and an old sheet to lay under her.

The area surrounding the bite on her face had swollen to almost golf-ball size. When we examined her, we saw the identifying marks of a snake bite on the right side of her lip.

Mikki undid Whiskers' collar so the swelling wouldn't make it too tight, and then she turned to me and said, "Get the prednisone."

We keep a few shots of prednisone on hand for emergencies like this, and though the shot has a few risks that go along with it, we felt it best to administer the prednisone to help with the effects of the venom.

After Mikki gave her the shot, I went out to look for the snake. It didn't take long for me to find it. Not twenty feet from the fence, I found a copperhead. I called my son, who came and caught it, then relocated it a few miles away. Below (or on the next page, depending on what device you're reading this on) is a picture of him holding it.

Jimmy holding copperhead

By about ten that night, the bite on Whiskers' face had swollen to the size of a tennis ball, and it was oozing pus and blood. We'd been through this already with Bear and another dog. We felt pretty sure that if we could keep the swelling down and get enough water in her, she'd probably be okay—assuming she survived the first night. If she did, she'd have a few tough days, but would likely come out of it.

Mikki stayed with Whiskers all night, gently applying ice-packed cloths and keeping the wound clean. Whiskers wouldn't let anyone else come near her. She wasn't a dog that allowed people to fuss over her, but she must have known this was serious, because she let Mikki do whatever she wanted that night.

In the morning things were no better; in fact, Whiskers looked worse. The swelling hadn't subsided and the wound looked terrible. More pus and blood oozed out and it had gotten a purplish hue. I called the vet, but he said at that point there wasn't much to do other than what we were doing.

Mikki nursed Whiskers as if she were her child. We continued doing everything we could to get her to drink water, and by mid afternoon, Whiskers showed a slight improvement. By midnight, the swelling had gone down quite a bit, though the wound was still oozing.

Every couple of hours, Mikki cleaned the wound and forced a little water into Whiskers. The next morning Whiskers went out to do her business, but she wouldn't come back inside. For all of her love and trouble, Mikki got a quick exit and a wag of Whiskers' tail as she walked off.

We tried coaxing her to come in, and then tried catching her, but it was no use. She made her way to a spot near the middle of the driveway and plopped down in the dirt.

To hell with healing—Whiskers was back on duty.

AN UNEXPECTED KISS

*L*ate one night while I was writing, I heard a noise outside. When I looked, I found Whiskers dragging herself toward the house. Her back legs were crippled and she was whimpering. I carried her inside and examined her but couldn't find anything wrong. No blood. No signs of a broken bone. And no indication of a snake bite. She'd already been bitten twice by copperheads, so we knew what to look for.

Mikki and I wrapped her in a blanket and laid her on the sofa. The next morning Whiskers still couldn't walk, so we took her to the vet. He didn't give us much hope, but we left her with him to see if he could do anything.

AFTER TWO WEEKS in the clinic she still couldn't walk. The vet said there was nothing we could do, and suggested putting her down. We decided to take her home instead.

For three more weeks we gave her pills and carried her out every day. There had been slight improvement, but not much. She couldn't walk ten feet without falling down. We decided we'd give it another few weeks.

Whiskers couldn't walk

The next morning around 6:30, I fed Dennis, our wild boar, fed the horse, gave Whiskers an anti-inflammatory pill and took her to the front yard. She could walk with help from me holding her back legs. After I got her to the grass, I went to the kitchen to make coffee. When I finished my coffee, I went back outside to get her—she was gone!

I looked everywhere and couldn't find her, so I called Mikki, and we both looked. Then we got the tractor and drove around the property —through the woods, around the pond... Whiskers was nowhere! I got a sick feeling in my gut. Something was wrong.

We started at square one. This time I walked every inch of the property, calling her name the entire time. After almost an hour, as I was making my way around the pond for a second time, I heard a whimper. I looked but couldn't see her. I called her name, and again I heard a tiny whimper. It was coming from the pond!

Our pond had been invaded by giant salvinia, a South American species of plant that takes over in a matter of weeks. It is damn near impossible to get rid of. When I got to the edge of the pond, all I could see was Whiskers' nose, and, when she bobbed her head, a bit of her mouth. She went under just as I got there.

Pond, showing giant salvinia

I jumped in and briefly went under, all the time I worried that Whiskers wouldn't be able to stay afloat. I finally reached her and grabbed hold of her neck. I pulled her to me and cradled her in my arms. While trying to make it to the shore, Whiskers struggled to stay afloat in my arms, and I struggled to stay on my feet, as the bottom of the pond put the definition of slippery to shame.

I managed to get Whiskers to the side of the bank and pushed her up on it, but she kept sliding back. The floor of the pond had a steep slope and I couldn't maintain my balance. I finally found a foothold on a branch and was able to stabilize my position. I gave Whiskers one big push, stabilized my position, and managed to crawl out onto the ground next to her. While I lay there on the bank with Whiskers, I leaned in close and said, "You damn crazy dog. You almost killed us both."

Whiskers, covered in mud

She let out a small whimper, and then she did something she had never done. Not once in the ten years I had her—she reached over and kissed me.

That may not seem like much for you people reading this. It's not much for any dog. But for Whiskers—it was huge!

Whiskers Had Never Kissed Anyone

- Not my son, when he carried her outside every day for a month after she'd been hit by a car.
- Not my wife, when she spent days tending to Whiskers after a copperhead bit her and her face swelled until she looked as if she had a grapefruit attached to it.
- Whiskers has never kissed my grandkids, my niece, or me. No one! Ever.

That kiss was magic! There's no doubt in my mind what it was. It was a thank-you kiss.

I took Whiskers back to the house, where we cleaned her and dried her off, and then we went about the arduous task of figuring out how she managed to escape the front yard. I had it fenced in with posts, corral board, and hog panel on top of that. It took us a couple of hours, but we finally found the spot where she had pushed the hog panel from the inside and managed to pop a few of the nails out. Once that was accomplished, she squeezed through the bottom opening.

WE SPENT two days re-attaching the hog panel to the corral board, making sure she couldn't escape again. Or so we thought.

VISITS FROM A FRIEND

*W*hiskers stayed in the house for several weeks after that escapade. We only let her out to go to the bathroom, and only then with supervision. Bear came to visit every day.

Ordinarily I would say he was simply coming to get Whiskers for their nightly date, but that wasn't the case. He would come during the day also, and he would sit in the house with her or lay in the front yard with her when we let her out. It was almost as if they were reminiscing about their old hunting days. It was touching to see a clear sign of affection, but it was also heartbreaking.

At that point I realized that even Bear was growing old. He no longer terrorized the neighborhood dogs or went searching for them on his nightly romps—though none were so bold as to walk down his street. And he wouldn't go out at night anymore. In the old days, if Whiskers wasn't available, Bear went alone.

He could still jump the five-foot fence on our property, and he still swam in the pond on hot days, but the signs of age were unmistakable. It didn't fully hit me until the second week of Whiskers' recovery.

We were in the front yard, and Bear had come to visit Whiskers. A UPS truck came down the driveway to deliver something—and yet Bear didn't even stir. Right then I knew he was getting old. In times past, he'd have jumped the fence and chased that truck down. One time he even jumped *into* the truck while it was moving. Bear would do anything to get another notch on his teeth. For whatever reason, he hated UPS and Fed-Ex trucks and their drivers.

For all that he hated delivery men or any adult not under his protection, he loved kids. He would tolerate almost anything from kids without so much as a growl. And he had a profound respect for the hierarchy of the pack. When Bear had first started coming by our place, about twelve years earlier, we had an older dog named Kelly. She was a feisty old girl who loved to fight and show she was boss.

Bear could have torn her apart if he wanted, but whenever Kelly tried to pick a fight with him, he simply avoided her. He'd step aside, like a matador dodging a charging bull, or he'd run a few steps to get away, and then let Kelly wear herself out. He also respected the other animals on the sanctuary—all but Dennis.

Dennis' story is too long to relate here; it deserves a book by itself, but the long and short of it is this. Dennis is a wild boar. He came to our sanctuary as a baby, and was almost dead. Now he's full grown and weighs almost four hundred pounds. Bear couldn't do anything to him if he wanted, but I have never doubted that he wanted to. And so did Whiskers.

I guess it was because of their run-ins with the boars in the woods. Bear and Whiskers don't bother the potbelly pigs or the feral pigs. Only Dennis seemed to irritate them. Dennis got under Whiskers' skin the worst. She was smart enough to know not to bother him, even though he was locked up behind a fence. But when the other dogs were out, she stirred them into a frenzy, inciting them to bother Dennis through the fence.

I figured that's what Whiskers and Bear must have been talking about during their sessions, sitting around telling tales like two old friends. I smiled as I watched them. They sure had tales to tell.

ONE LAST HURRAH

*T*he next week carried with it the promise of hope. Whiskers seemed to be walking better, although she still had a bad limp and a crooked gait. By Wednesday, she was adamant about going out at night. We didn't let her, and the next night was even worse. Then I found out why.

Biscotti had been going out at night by herself, and though she was restricted to the fenced-in area, it was still seven acres. It was during those times that Whiskers became the most vocal in her demands. If any of us dared go near the door, she tried squeezing out. I went for a walk, carrying a big flashlight and discovered the reason—Biscotti had treed a raccoon, and she was barking like crazy. I don't know how Whiskers knew this was an important bark, but she did, and it was driving her crazy.

View of pond from balcony

I got Biscotti inside, but the next night was a repeat performance. On the following night, we kept Biscotti in. Still Whiskers threw a fit, and she continued her protests for two more days.

On Monday, I got up early and let Whiskers out for her morning business. I made coffee. Half an hour later I went out front to get her, but she was nowhere to be found. I soon discovered a hole under the gate where she had gotten out, and before long I found Whiskers. She was leading the pack on what appeared to be a tour of the property. Biscotti, Mollie, and even Briella were following her around as if she were the tour guide.

For the next few days I observed Whiskers taking them new places each day. The odd thing was, for the first time in her life she seemed content being inside the fence. It was something I thought I'd never see. They chased a nutria, stalked squirrels, and even rousted a couple of rabbits. Late that night, they all went wild when the coyotes sounded off.

Bear hadn't come down for a few days, but it didn't surprise me; he seemed to have aged a lot in the past few months, and he no longer went out hunting at night.

The next morning was the same routine: I got up, let Whiskers out, made coffee, and then went to watch her. I didn't see her at first, but then I heard a ruckus at the other end of the property.

I got down there as fast as I could, and found Whiskers leading three other dogs in a charge against two raccoons. I grabbed a tree branch to try and stop them, but the dogs got one of the raccoons before I could make them to stop. The other raccoon got away, but both the dogs and the raccoons had done damage. Biscotti needed stitches in her ear; Mollie's eye was bleeding badly; and Corny's throat was close to needing a few stitches. Whiskers—nothing. Not a scratch. At least it seemed that way.

Biscotti's ear

By nightfall, Whiskers couldn't walk again. By morning she couldn't move any of her legs. We rushed her to the vet and, after extensive testing, he determined she had major spinal damage. She couldn't feel anything below her neck.

A GOODNIGHT KISS

*D*espite trying everything possible to get the doctor to say he could fix her—he simply couldn't, so we prepared ourselves for the inevitable. My wife and I had been through this too many times. For all the wonderful benefits of running an animal sanctuary, having to say goodbye is one of the few horrible consequences.

The doctor left the room, and we spent some time with our magnificent dog. My wife combed Whiskers' hair, perhaps the first time we ever got to do that, and we talked to Whiskers like old friends.

This may have been the most difficult time we'd ever had letting go because, mentally, Whiskers was alert. And so *damned* alive. I couldn't stop petting her, and talking to her, but we knew it was time.

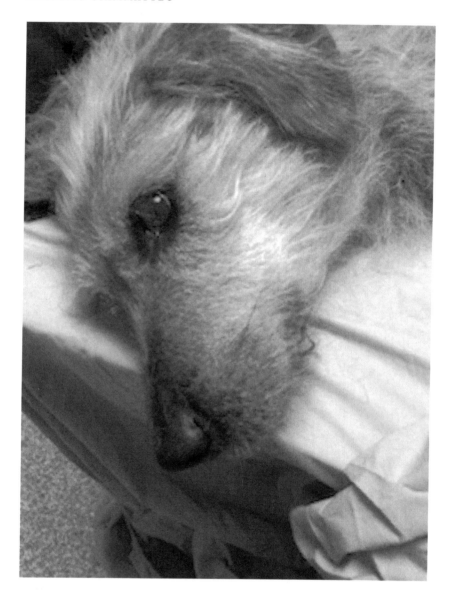

I leaned down, gave her one last kiss, and whispered my *goodnight message.*

Mikki did the same.

We didn't give her a goodbye kiss; that was far too permanent.

Instead, we gave her a *kiss goodnight*, and, we promised to meet her on the other side.

DIG THE HOLE DEEP

oe and I were down to working with shovels now. We'd gone as deep as we could with the tractor, but we still had farther to go. Mikki insisted on digging the holes deep. She worried that the coyotes would dig up the remains. That wouldn't do for any animal, but it was especially important with Whiskers. She'd spent ten years fending off coyotes. The least we could do was make sure none got her now.

Mikki lifted her blanket, gave her another kiss, and then we lowered Whiskers into the hole. Mikki climbed in and set the pillow under her head. I almost laughed. It may have been the first time in her life the dog had rested her head on a pillow. I guess it was a fitting time for it.

For me, the most difficult part came next—covering her up. As much as I disliked digging the holes, it tore at my guts to cover them. I waited a few seconds, then got in the tractor, took a deep breath, and lowered a scoop of dirt on top of her. And then another. And another.

That night, as Mikki and I watched TV, I realized Biscotti wasn't with us. She was normally curled up beside me on the couch. "Where's Biscotti?" I said.

Mikki looked around and checked behind the sofa. "She's not here." Then she panicked. "Mollie's not here either."

I got nervous, worry creeping through my bones. I got the flashlight, and we went outside. I called Biscotti's name a few times, then Mollie's, but we got no response. As we walked toward the pond, Mikki tugged on my arm. "Look by the drive," she said, and her voice cracked. "They're sitting by Whiskers."

When I looked over, I almost cried. "There she is," I said to Mikki.

We walked over to where we had buried Whiskers—by where she guarded the property every night—and there sat Biscotti and Mollie, not two feet from Whiskers' grave. Biscotti's head was raised in the air, and she was staring off into the woods.

Biscotti and Mollie

"I guess we're safe again," Mikki said. "Whiskers trained her replacement."

A NEW GUARDIAN

*a*t nights, while the other dogs are sleeping on the couch or lying in their beds, I hear Biscotti doing her job. She doesn't have the hoot-owl kind of bark that Whiskers did, but Biscotti patrols the same spots, and she won't let anything come onto the property.

I'm convinced now that those last few days—when Whiskers seemed to have gotten a new sense of purpose—it wasn't a resurgence of energy or life. She was just hanging on long enough to show Biscotti what she had to do. A master training her apprentice.

It was a cool autumn night when I finished this story. The moon was full, and the woods were filled with noise. I walked outside and strolled through the front property.

Everything seemed in place: The frogs were chirping, flying squirrels were chittering away, crickets sounded from over by the barn, and I knew in a few minutes that the 11:00 train would draw howls from the coyotes. Everything seemed perfect...but I would have given *anything* to hear Whiskers' bark just one more time.

MY GOODBYE MESSAGE TO
WHISKERS

*W*e've been together a long time, Whisk. You made us angry. Made us worry. You've even made us cry. But you made life exciting, and you gave us a lot of laughs. Best of all, you made us love you.

I don't know if you understand what's going on. Or if you're scared. But don't worry, you're going to a place like you've never seen.

Pretty soon your eyes will close. When they open again, you will be with your friends. Slick will be there, and you can play with him as you did when you were puppies. Princess will teach you the ropes and show you the new territory. Kelly will be there to fight with, and no one will get hurt.

And before you know it—while you're having fun—your best friend, Bear, will show up. And the two of you can roam the woods forever.

EPILOGUE

*T*he sanctuary is full of stories. This is just one of them. For us to keep going, we need your help. We have forty-one animals and many of them are growing old; in fact, Shinobi, the potbelly pig who started it all turned twenty-four in July, which was closing in on the world record.

She passed away in September, 2016, but she went out like a trooper.

The bottom line is that as these animals age, especially with some of the problems they have, medical costs skyrocket, as does the cost of repairing fences, barns, etc. Feed bills lower somewhat, but it's almost nothing when compared to the increase in other costs.

For many years we were an IRS-sanctioned charity, so gifts were tax deductible; now we have forfeited that, as my own health issues did not allow us the time to keep up with required paperwork. I'm sure not being sanctioned by the IRS will curtail donations, and I understand, so know that if you decide to donate, it *is not* deductible.

I know everyone has favorite charities. All I'm saying is that if you donate toward Tuskany Falls, you can be assured that *all* of the money goes to helping the animals. Not one cent goes to anything else—not to administration, or salaries, or *miscellaneous* costs. All of it goes for feed, healthcare, or repairs.

Note:

You can even help without donating, and it's as easy as can be. Simply go here, before you buy anything from any store, then click the links for the appropriate store listed below the books at the top. Our sanctuary will get credit for the proper affiliate code if you do. You don't have to buy a *book*; buying anything will suffice.

So, how about it? Can you afford to skip a few cups of coffee for someone like:

Petey, aka. Sweet Pete

My granddaughter, Adalina, and her best buddy, Joe the Horse

Bear lounging in the grass

Bear walking up from the pond

Giant Briella. She was 190 pounds

A coyote, mortal enemy of Whiskers

Dennis, the great and wonderful

Whiskers after, an obviously, dirty adventure

Joe the horse sharing food with the pigs

Five feral kitties, who now have a home here

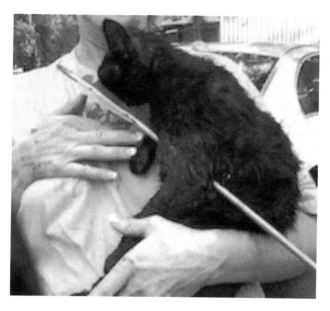

Hotshot—my wife found him on the road. He'd been shot with an arrow.

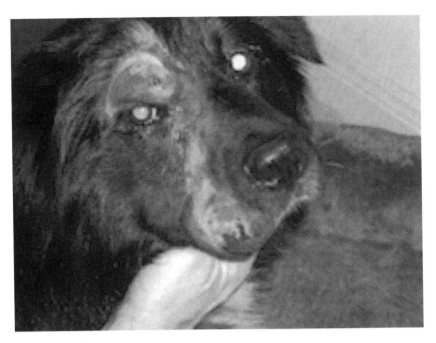

Bear, after his raccoon attack

Queen Shinobi, as she was known. She started it all.

AFTERWORD

Regarding Communication

I have nothing scientific to offer, just the observations of someone who has taken care of dozens and dozens of animals of all kinds. I know that many others have done the same, but as a thought, perhaps this was different because of the circumstances.

This wasn't the normal way a dog, or even a group of dogs, is raised. These dogs were a pack. They weren't under my thumb, ruled by a human master. They, or should I say Bear and Whiskers, were in charge. The dogs did as they pleased, roaming through acres of wooded property, filled with wildlife.

Bear and Whiskers had free roam of everything and fended for themselves. I think that may have been the difference. They didn't live to please us; they lived for themselves, and while Whiskers and Bear listened to us (to a degree), I think all of the dogs listened to Bear more.

Regardless of what it was, or is, there's one thing I'm certain of—these dogs *spoke* to each other, and they *understood* each other.

And all I can say is I'm glad I was part of it and got to see the interaction and be a part of the magic.

ABOUT THE AUTHOR

Giacomo Giammatteo is the author of gritty crime dramas about murder, mystery, and family. He also writes nonfiction books, including the No Mistakes Careers, No Mistakes Publishing, No Mistakes Grammar, and No Mistakes Writing series.

When Giacomo isn't writing, he's helping his wife take care of the animals on their sanctuary. At last count they had 45 animals—11 dogs, a horse, 6 cats, and 26 pigs.

Oh, and one crazy—and very large—wild boar, who takes walks with Giacomo every day and happens to also be his best buddy.

nomistakespublishing.com
gg@giacomog.com

ALSO BY GIACOMO GIAMMATTEO

You can see all of my books here.

And you can buy them on the platform of your choice.

Nonfiction :

No Mistakes Resumes, Book I of No Mistakes Careers

No Mistakes Interviews, Book II of No Mistakes Careers

Misused Words, No Mistakes Grammar, Volume I

Misused Words for Business, No Mistakes Grammar, Volume II

More Misused Words, No Mistakes Grammar, Volume III

No Mistakes Writing, Volume I—Writing Shortcuts

How to Publish an eBook, No Mistakes Publishing, Volume I

How to Format an eBook, No Mistakes Publishing, Volume II

eBook Distribution, No Mistakes Publishing, Volume III

Uneducated

Fiction:

Friendship & Honor Series:

Murder Takes Time

Murder Has Consequences

Murder Takes Patience

Murder Is Invisible

Blood Flows South Series:

A Bullet For Carlos: A Connie Gianelli Mystery

Finding Family, a Novella

A Bullet From Dominic

Redemption Series:

Necessary Decisions: A Gino Cataldi Mystery

Old Wounds

Promises Kept, the Story of Number Two

Premeditated

~

OTHER BOOKS COMING SOON:

You can always see the current and coming-soon books on my website.

Fiction:

A Promise of Vengeance (Fantasy)

My first fantasy, and the first book in a four-book series—the Rules of Vengeance. (Three are already written and the fourth is being outlined.)

A Hard Life, the Story of Tip Denton

Memories for Sale (mystery/sf)

The Joshua Citadel (SF novella)

Nonfiction:

Whiskers and Bear—Volume I of the Life on the Farm Series (sent to editor)

No Mistakes Writing, How to Write a Bestseller

Children's Books:

No Mistakes Grammar for Kids, Volume I—Much and Many (Sent to editor)

No Mistakes Grammar for Kids, Volume II—Lie and Lay (Sent to editor)

No Mistakes Grammar for Kids, Volume III—Then and Than (Sent to editor)

Shinobi Goes to School—Life on the Farm for kids. (working on illustrations)

Get on the mailing list and you'll be sure to be notified of release dates and sales.

Mailing list

And don't forget to leave a review!

To leave a review, go to the site, then click on the retailer of your choice.

PART I
CAST OF CHARACTERS

The Life on the Farm series consist of stories featuring animals from our sanctuary. The stories are based on the *personality* of the animals as we see them.

Obviously animals can't talk, at least not in human language, so while they can't tell us specifically what they want or need, at times it's almost easier to understand what they are trying to say.

Without further delay, I'll introduce you to the animals. Be patient, as there are a lot of them.

PIGS

Shinobi—Sometimes known as Queen Shinobi. She is the one who started it all. She was the first animal rescue we had, and as of last September (2016) she either was the oldest pig in the world or the second oldest.

Queen Shinobi

She turned twenty-four in July, 2016. She passed away in late September, but her indomitable spirit and winning personality will be with us forever.

Starbuck—Shinobi's sidekick. Starbuck was the second pig on the sanctuary.

Starbuck

The Squeaks

Mama Squeak—Squeak came to us on one of the hottest days of July, in 2006

She eventually had six piglets, which we named "Squeakers" when referring to them as a bunch. Of course, they all had individual names. I've listed them below.

Marco—One of the Squeakers

Punch—The star of many of the stories.

Willow—The shy girl.

Bertie—AKA Bertie the Bully.

Speckles—Good-natured Speckles.

Inky—Always happy.

Wild Bunch plus Angel

Fiona—Sadly, Fiona is gone now, but she had a magnificent life. We rescued her and her babies from a wooded section off the interstate north of town. Her babies are listed below. We called them the Wild Bunch.

Morgan—the tough guy.

Clyde—Mr. Speed.

Bonnie—The quickest.

Juice—Sweet Juice the screamer.

Paige—The protector.

OTHERS

Coco

Cool Pig

Piccolo

Angel and Lucy

Oprah

Pearl

Giacomo

Nina

Petey, aka, Sweet Pete

Spike

Lucy

Angel, letting me know she's hungry

Lanie

And of course, Dennis, who gets his own page (included in the author bio).

DOGS

Bear being lazy

Bear wearing glasses

Baby Biscotti

Big Briella

Butters with her favorite toy

Slick with his favorite friend

Pretty Girl, a pit bull rescue we had

Farrah, a blue-merle Australian Shepherd

Mollie sitting alongside Kirby

Lucci and Vinnie

Sandi, eating the wall

Slick, with his buddy Louise

Molli on the couch with my niece, Emiliana

Kelly

Flash, snarling

Kassie

Candy

Whiskers, one of the best

Princess

CATS

Mamma Cat and all her babies

Mama Cat—Manx (no tail)

Hiss—son (tail)

Boing—son (tail)

Carmel—daughter (no tail)

Shy Girl—daughter (tail)

Baby Bats—daughter (no tail)

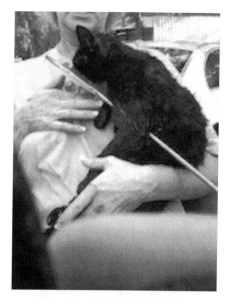

HotShot—our rescue that had been shot with an arrow

Meowie, another feral we found

All kittens eating

Four cats on windowsill

HORSE

Joe

Adalina kissing Joe

DENNIS

Dennis the great and wonderful

Made in the USA
San Bernardino, CA
03 December 2018